Examining
Biofuels

Chase Robertson

CLARA
HOUSE
BOOKS

First published in 2013 by Clara House Books, an imprint of The Oliver Press, Inc.

Copyright © 2013 CBM LLC

Clara House Books
5707 West 36th Street
Minneapolis, MN 55416
USA

Produced by Red Line Editorial

The publisher would like to thank D. Leith Nye, Education and Outreach Specialist for the Great Lakes Bioenergy Research Center, for serving as a content consultant for this book. The publisher also thanks Ralph Lerner, Edeniq, Inc., Visalia, California, for his insights.

Picture Credits
Shutterstock Images, cover, 1, 14, 24–25, 33, 39, 41; Vasyl Dudenko/Shutterstock Images, 5; Library of Congress, 9; Red Line Editorial, 10, 19; Melinda Fawver/Shutterstock Images, 12; Daniel Yordanov/Shutterstock Images, 16; Inacio Pires/Fotolia, 18; LiteChoices/Shutterstock Images, 21; Fotolia, 22; Eric Tadsen/iStockphoto, 28; Monkey Business Images/Shutterstock Images, 31; Carolina K. Smith, M.D./Shutterstock Images, 34; Darren Baker/Shutterstock Images, 37; Adrian Britton/Shutterstock Images, 45

Library of Congress Cataloging-in-Publication Data
Robertson, Chase.
  Examining biofuels / Chase Robertson.
    pages cm. -- (Examining energy)
  Audience: Grades 7 to 8.
  Includes bibliographical references and index.
  ISBN 978-1-934545-40-9 (alk. paper)
  1. Biomass energy--Juvenile literature. I. Title.

  TP339.R63 2013
  662'.88--dc23

                        2012035242

Printed in the United States of America
CGI012013

www.oliverpress.com

Contents

CHAPTER 1

Green Energy

Have your parents ever complained about the high cost of gasoline? What about the high cost of electricity? People all over the world use a lot of energy. Right now, much of the world's energy comes from non-renewable sources. These non-renewable sources, such as oil, can harm the environment, and they will eventually run out. With the human population growing across the planet, the cost to produce clean energy for the world also rises.

Scientists are constantly looking for ways to improve our sources of energy. They want to find ways of producing energy that are more efficient, less expensive, and better for the environment than our current energy sources. Alternative energy research focuses on balancing our energy consumption and production costs against what's best for our environment. One possible source of energy might come from biofuels. Finding a way to make biofuels a practical energy source for transportation could help to keep the environment cleaner.

Biomass such as corn could play a role in powering our future.

It could also reduce the country's need to import fuel from overseas. But developing this alternative fuel source requires cutting-edge research.

Biofuels are made from organic material, or living things, also known as biomass. Burning organic plant material releases the energy stored during photosynthesis, the process of turning sunlight and oxygen into food. Biofuels aren't the only fuels made from organic material. Petroleum-based fuels such as gasoline and diesel are made from organic material buried underground for millions of years. Because these fuels take so long to make, we call them fossil fuels and consider them to be non-renewable.

Scientists are interested in biofuels because they take much less time to make than fossil fuels. Biofuels are liquid fuels for transportation made from renewable organic material—we can always create more biofuels. The production of biofuels for use in transportation could be an important way to help us decrease our dependence on fossil fuels.

EXPLORING BIOFUELS

In this book, your job is to learn about biofuels and their place in our energy future. Where did biofuels come from? How do they work? Which fuel provides the most energy? Can we produce enough economically to meet the needs of the world? What is the cost to the environment?

Esmeralda Sanchez is researching biofuels for a middle-school science project. She is meeting with energy experts and farmers to learn more about biofuel's role in the future of energy. Reading her journal will help you in your own research.

CHAPTER 2

Biofuel Basics

My first stop is my science teacher's room at my middle school. Inside Mrs. Baker's lab, I see tables covered with corn stalks, long grasses, wood chips, yard clippings, newspaper, a rotting banana peel, and even a piece of dried cow manure. The word biomass blinks on the interactive whiteboard.

"Biomass is organic material made from plants and animals," Mrs. Baker says. She explains that all of the objects on the tables are organic material. Organic material stores and provides energy in an endless cycle. Plants and animals give off carbon dioxide. During photosynthesis, plants use the sun's energy and carbon dioxide to make food for themselves, emitting oxygen. Animals breathe the oxygen and eat the plants. The animals breathe out carbon dioxide, which the plants absorb and turn into energy during photosynthesis, thus starting the cycle over again. Biomass stores some of this energy, which can be used as food for humans and animals. It can also fuel campfires, cars, and other human energy needs.

"Take a look in your trash barrels at home," Mrs. Baker says. "Every day you throw away biomass that could be used to create energy. Since biomass has the unique ability to become liquid fuel, we can use that wasted energy to help decrease our dependence on non-renewable energy sources, such as oil."

UP IN SMOKE

Do you have a fireplace in your house? Wood is a common form of biomass. When wood burns in your fireplace, it releases energy in the form of heat. Burning wood to generate energy emits more carbon into the air than fossil fuels like gasoline. However, unlike gasoline, the carbon released by burning wood is already part of the carbon cycle, so no new carbon is released. Still, a wood-burning heating system is not an ideal solution. Wood smoke contains harmful pollutants that can be bad for the environment. Sometimes countries cut wood faster than it grows, which leads to a decline in the number of trees in our forests. Fewer trees means less absorption of carbon through photosynthesis, but planting new trees can replenish the forests and help the environment.

Mrs. Baker explains that biofuels, fuels made from biomass, have been around for a long time. When Henry Ford sold his first Model T car in 1908, he planned to fuel it with ethanol, a type of alcohol made from corn or other energy-producing crops. Early diesel engines ran on peanut oil, a kind of biomass, rather than gasoline. In fact, gasoline was a waste substance left over from the production of kerosene, a type of oil used in lanterns before electricity was common. Ford's Model T changed

Ford's Model T car was originally designed to run on ethanol.

everything. Gasoline became the fuel of choice because it was cheap and released a burst of energy when it burned.

Fossil fuels such as petroleum that are used to make gasoline are ancient biomass. They are made from plants and animals buried underground and cooked by the Earth's heat and pressure over millions of years. Because fossil fuels take so long to make, we use them up more quickly than they can be created. "Petroleum is becoming more expensive because the world is using more of it," Mrs. Baker says. "Besides, digging fossil fuels from the ground and burning them sends extra carbon elements into the environment in the form of carbon dioxide."

Carbon is an element common in organic matter. Mrs. Baker flashes an illustration of the carbon cycle on her whiteboard. We know from the news that extra carbon can

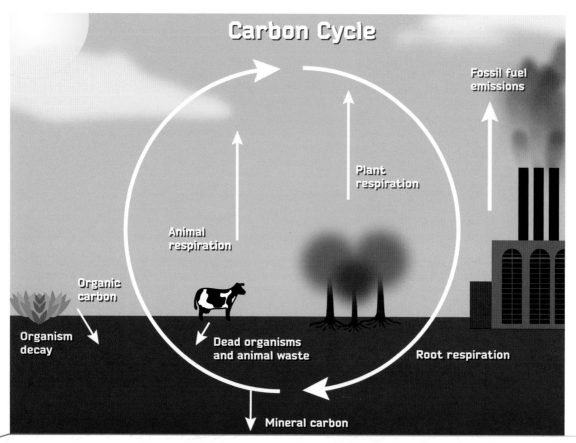

TIPPING THE CARBON BALANCE

Carbon is part of all living things and is always on the move. In the carbon cycle, plants absorb carbon through photosynthesis. Carbon moves from plants to animals through the food chain. Every time animals exhale and plants respire, carbon moves into the air. When plants and animals die, carbon moves into the ground. Oceans absorb carbon, too. By burning fossil fuels, we create extra carbon in the cycle; plants and animals cannot consume the carbon as fast as it is being added. This leads to an excess of carbon in the atmosphere and oceans, which can pollute our environment.

disturb the balance of Earth's delicate carbon cycle. Some people believe that the extra carbon in the air, primarily in the form of carbon dioxide, can create climate change, which may affect weather patterns.

"Today's biofuels speed up the time it takes to produce fuel," Mrs. Baker says. "Instead of waiting millions of years, modern biofuel can be made in months."

"If biofuels are so great, why haven't they replaced gasoline?" I ask.

Mrs. Baker explains, "Gasoline contains more energy per gallon than biofuel, so it's more efficient."

"Meaning we get more miles per gallon when we use gasoline in our cars, right?" I say.

"Right," she says. "Changes in engine technology could make biofuels almost as efficient as gasoline one day. Still, testing biofuels in a lab is quite different from making enough to power an energy-hungry world. We have to learn how to manufacture enough biofuel for the whole world."

Mrs. Baker continues, "Third, biofuels currently cost more than gasoline. To make them cost effective, we need to use existing oil refineries, trucks, and gas pumps. Scientists are trying to create biofuels that can fit into our current energy production and can be used without too much extra expense."

Gasoline powers most cars today—but gasoline releases carbon dioxide and other pollutants into the atmosphere.

I thank Mrs. Baker and say good-bye. I still have questions about biofuels. Mrs. Baker said I throw away biomass energy every day, which got me thinking. My parents recycle paper and plastic, and compost yard clippings, but a garbage truck still empties our garbage can every week. Where does that garbage go?

CHAPTER 3

From Garbage to Energy

I head to my town's landfill where I meet with Steve Miller, the landfill director. Steve tells me that if I am like most Americans, I throw away approximately 4.4 pounds (2 kg) of trash every day. He says, "Some of that trash is recycled and composted, but the rest comes here."

He gestures to a mountain of garbage. Trucks dump enormous loads of garbage onto the pile. Tractors spread out the garbage. More dump trucks cover it with dirt.

"Bacteria live in landfills," Steve says. "These bacteria decompose paper, wood, food scraps, and other biomass to make biogas. The gas contains methane."

Steve leads us to a grass-covered section of the landfill. It looks almost like a park, except for the pipes sticking out of the ground.

"Methane is energy rich," he says. "We use the energy created from burning it to heat and cool our offices at the landfill. We have enough energy left over to sell to the power company to provide electricity to approximately 8,000 homes."

Steve's comments make me think about my uncle who owns a dairy farm. He shovels cow manure into a big tank he calls a digester. I ask Steve if the digester works the same way as the landfill.

"A digester is a bit like your stomach," Steve says. "In your stomach, enzymes break down the food you eat into sugars that your body uses for fuel. In a digester, anaerobic bacteria break down the manure, which

CARS THAT GO CLUCK–CLUCK?

People all over the world love to eat chicken. All that chicken generates about 11 billion pounds of poultry guts, bones, blood, and feathers each year. Feathers become pillow stuffing. The rest of the chicken usually becomes low-quality animal feed. What if we could cook all that waste to make biofuel? Researchers at the Renewable Energy Center at the University of Nevada in Reno found that leftover chicken parts contain approximately 12 percent fat. Fat is an essential ingredient in biodiesel, a more environmentally friendly alternative to diesel fuel. Chicken parts and other plant and animal matter would not be able to meet all of our fuel needs. Still, researchers estimate that they can produce about 593 million gallons (2 billion L) of biodiesel each year from the billions of pounds of chicken waste the poultry industry produces.

Most Americans' trash ends up in a landfill.

A lot of the trash in landfills is made up of biomass.

releases biogas that can be used to generate electricity for your uncle's farm."

"My science teacher said biofuels could help us decrease our need for oil," I tell him. "Do you make liquid fuel here?"

Steve shakes his head. "Only electricity here. But your science teacher is right, and that is what makes biofuels special. Natural gas is also made primarily of methane and can be used to produce electricity. But biogas is produced renewably and natural gas is not. Wind, solar power, and nuclear energy also make electricity, but biomass has the ability to produce liquid fuel that would be convenient for transport, just like gasoline."

CHAPTER 4

Corn: Food or Fuel?

I remember the banana peel on Mrs. Baker's lab table. Food is biomass, too. I decide to check out an Iowa cornfield. I ride past miles and miles of cornfields occasionally dotted by a farmhouse. Before I know it, I'm standing with farmer Trent Gable in the middle of his cornfield. The stalks are higher than our heads. Fat ears with golden corn silk grow on each stalk. They look tasty, but Trent sells his corn for fuel instead of food.

"Corn has fed the world for centuries," he says. "Now it's fueling the world with ethanol. We use the same planting and harvesting techniques whether our corn grain becomes food or ethanol. First we plant the corn seeds. Next we fertilize and water them. Then we harvest them."

Trent explains that ethanol is the only biofuel currently available on a large scale. Right now, ethanol accounts for approximately 10 percent of the U.S. gasoline supply. Essentially, ethanol is the same kind of alcohol as in the beer or liquor adults drink. When this alcohol is mixed with gasoline, though, it helps reduce greenhouse gases.

I look at the ears of corn all around me. "Do you make the ethanol here on the farm?"

Trent shakes his head. "We harvest the corn and grind up the kernels. Then we ship the corn to a factory. Corn is usually more than 70 percent starch, a type of carbohydrate. The factory converts the starch to sugars using enzymes."

Corn can be used to feed humans and animals or to create the ethanol that helps fuel cars.

I remember my landfill visit where Steve explained how enzymes break food into sugar.

"Once the corn starch has been broken down, tiny organisms called microbes eat the sugars," he says. "These microbes, such as yeast or E. coli, produce ethanol as they munch away."

"Why can't we just fill our gas tanks with ethanol?" I ask.

"Like gasoline, ethanol is not a perfect fuel," Trent says. "First of all, the energy content is low. You need about one-and-one-half gallons (5.7 L) of ethanol to equal the energy power of one gallon (3.8 L) of gasoline. Second, it mixes too easily with water."

I already know that petroleum is an oil-based product. I remember from science class that oil and water do not mix. But I don't understand why ethanol mixing with water is bad.

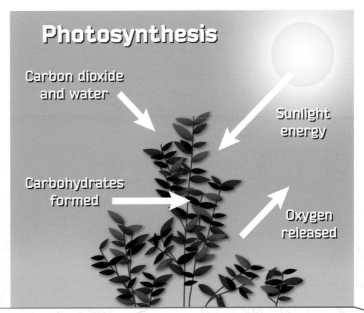

PHOTOSYNTHETIC BALANCE

Have you ever heard that talking to your plants makes them healthy? Although plants may not be able to hear you, this is true because as you breathe out you expel carbon dioxide. Plants use your carbon dioxide, along with water and sunlight, to make carbohydrates. In turn, the plant gives off oxygen, which you breathe.

"First, water doesn't burn," Trent says. "Additionally, water messes up the way engines work and weakens the energy content of the fuel. And because ethanol mixes with water so easily, it doesn't travel through pipelines well."

Trent explains that oil is transported from oil fields to refineries in vast pipelines. We can't transport ethanol through these pipes because small traces of water in the pipes mix with the fuel and reduce its effectiveness. An important goal for renewable energy is making the new energy work with the energy we are already using. Because we can't use the existing oil pipelines, we would have to find a new way to move the ethanol, which would be more expensive.

Additionally, some people worry about what

SWEET FUEL

During the 1970s, the world faced an energy crisis. As oil usage increased, the demand for oil became greater than the supply. Prices skyrocketed and led to an energy shortage worldwide. As a result, Brazil decided to reduce its oil consumption by mixing its gasoline with ethanol. Instead of using corn to make the ethanol, Brazilians used one of their most common crops— sugarcane. Sugarcane contains sucrose sugar in its stalk. Sucrose is the type of sugar you might find on your table. Since sugar is one of the favorite foods of certain ethanol-producing microbes, sugarcane might be an even better way to create ethanol than corn. Corn must be replanted every year, but sugarcane only needs to be replanted every six years. Today, approximately 90 percent of Brazilian vehicles run on a mixture of gasoline and ethanol. Brazil estimates that it has avoided the production of 660 million tons (600 million metric tons) of polluting carbon dioxide.

Sugarcane is an important crop in many parts of the world. In some countries, it is used to create ethanol.

will happen to food production if more and more farmers use their fields to plant crops for ethanol rather than for food. If the demand for corn for both food and fuel goes up, corn prices will also go up. This can affect other food prices. Cattle often eat corn. If farmers have to pay more for the corn to feed their cattle, the price of beef might also go up.

Corn is an important part of the U.S. economy. But the price of corn grain doesn't just affect the price of ethanol; it also affects the price of food.

Trent adds, "It might not be the perfect fuel, but ethanol does help to reduce your family's energy costs. In general, ethanol sells for about two-thirds the price of gasoline. At $4.00 per gallon for gasoline, ethanol would cost a little more than $2.50."

"Anything else?" I ask.

"Yes, ethanol-gasoline blends have become very common now, but not all ethanol is made from corn," he adds. "Ethanol made from sugarcane, for example, is actually the cheapest source of ethanol."

I thank Trent for teaching me more about ethanol. Next stop: a switchgrass field in Louisiana.

CHAPTER 5

Energy Grass

Today I am visiting switchgrass farmer Sarah Rickles at her field in Louisiana. As I get close, I notice long grass bordering both sides of the road. It is much taller than I am!

Sarah greets me with a warm smile. "Hi, Esmeralda. Welcome to our switchgrass field."

"What's switchgrass again?" I ask.

"We call it a purpose-grown grass because it is not part of the food chain," Sarah says. "This means that it doesn't compete with our fuel supply. All of the grass on our farm is used to make biofuel. Switchgrass is ideal because it grows on just about any type of land. It doesn't need prime farmland with excellent soil the way corn does. Another bonus: our farm uses little to no fertilizer, and our crop is relatively drought tolerant, so we don't use that much water. Switchgrass grows back after it's harvested, too. We don't need to replant it every year."

The grass grows in rows like corn, but the rows are much closer together. Every so often, I see a path between the rows. The tire tracks on the path tell me that a harvesting machine drives down these paths.

"Switchgrass produces advanced biofuel," Sarah says. "It is full of cellulose, a type of carbohydrate." I remember Trent's field, where I learned that enzymes can convert carbohydrates into sugars. When microbes eat the sugars, they create ethanol.

Switchgrass is native to the United States and naturally grows on prairies.

Sarah continues, "The high cellulose content of the grass produces more fuel per acre than corn grain. One acre of corn can produce about 400 gallons (1,500 L) of fuel, but an acre of switchgrass can produce about 1,000 gallons (3,800 L) of fuel."

"Is switchgrass the only grass that will work?" I ask.

"Good question," Sarah says. "That depends on where you live. Switchgrass grows well in the eastern, southern, and central parts of the United States. Miscanthus is another plant with high cellulose content. It is a native plant of Southeast Asia, so it grows well in warm climates. We can also use the cellulose from corn stover, or the leaves and stalk."

CHEMICAL SOUP

Carbon is the key ingredient in the creation of biofuels. Ethanol is a compound, or a substance made of two or more elements. Ethanol contains approximately 70 percent of the energy in a gallon of gasoline. Butanol is a 4-carbon compound made from biomass. Both of these compounds contain oxygen. In the world of biofuels, more oxygen equates to a lower energy value. Compounds that are made of only carbon and hydrogen are known as hydrocarbons. These are the building blocks of petroleum-based fuels. Scientists are working to make biofuels mimic the properties of gasoline and diesel for a more flexible and efficient renewable fuel. Grasses could not only power our cars, but also power jet engines, buses, and the farm equipment used to grow the biomass.

Sarah also explains that scientists are experimenting with fast-growing trees such as poplar and willow. These trees grow back after harvesting, but right now they cost more to grow than switchgrass.

The tips of the switchgrass sway in the breeze. A picture forms in my head of my dad stuffing blades of switchgrass into his gas tank. I know carbohydrates become sugars and sugars become ethanol, but I want to know more about how switchgrass turns into the fuel that could power my dad's car.

CHAPTER 6

From Grass to Gas

In the morning, Sarah takes me to an ethanol demonstration plant. The plant makes ethanol, but it does not sell it commercially. We drive past rows and rows of switchgrass before stopping in front of a factory where a truck dumps a load of cut grass into a waiting container.

She introduces me to the plant manager, Michael Singh, who hands me a hard hat. "You're just in time to make a batch of ethanol," he says.

Sarah and I follow him into the building. My sneakers squeak on the polished cement floor. Michael talks to me as we walk.

"Cellulose gives switchgrass its structure, so it is tough stuff," he says. "The first thing we must do is blow apart the cell wall to get to the sugar. The sugar in cellulosic plants is not as easy to get to as in corn grain."

"There are several ways to do that." Michael points to a stainless steel tank. "We use a biochemical process at this factory. First, we pre-treat the grass with an organic mixture that starts dissolving the tough cell wall—similar to the way laundry detergent dissolves stains on your clothes. Then our specially designed enzymes separate the cellulose into individual glucose or sugar molecules. A molecule is the smallest unit of a substance that still has the properties of that substance. Microbes eat the glucose molecules and make ethanol. This is called fermentation. After producing the butanol and ethanol fuels, we're left with solid plant waste that we use to power the factory."

I explain to Michael that I visited a landfill and know how garbage becomes electricity. As we walk through the factory, I notice shiny pipes snaking between tall tanks. Computer monitors flash temperature readings, warnings, and graphs.

Michael raises his voice over the sound of the machines. "Think about a car's engine. Essentially, gasoline mixes with air in the engine and ignites to release energy. Some ethanol plants heat biomass to release energy. This is called gasification."

Michael tells us that this process of heating biomass mixes in a small amount of air. This mixture produces gases, tars, liquids, and charcoal. Scientists weed out the molecules they

Ethanol refineries convert switchgrass, corn grain, and other biomass into fuel.

don't want and use chemical catalysts to speed up the creation of hydrocarbons.

Unfortunately, Michael says there are also problems with this process. "The factories require lots of money to build and maintain. The costs are so high, in fact, that it's too expensive to produce ethanol from switchgrass on a large scale, at least right now."

"But there is another way to blow apart the cell wall and get at the sugars. We can heat biomass without air," Michael says. "This method is called pyrolysis. In pyrolysis, we heat the biomass until gases are produced. Because there is no oxygen, the gases can't ignite. When the gases are cooled, they condense to create liquids. The liquids are bio-oils. The trick is to discover the best temperature and to choose the correct conditions to speed up the reaction. The good news is that bio-oil has similar properties to gasoline, and it's cheap."

"And the bad news?" I ask, because I've learned that no energy source is perfect.

BP BLAZES A TRAIL

British Petroleum (BP) hopes to open a cellulosic ethanol factory in 2014. Located in Highlands, Florida, the farm will grow 20,000 acres (8,000 ha) of purpose-grown grass to produce about 36 million gallons (136 million L) of fuel. That is equal to 20,000 football fields of grasses that can reach more than 10 feet (3 m) high! BP will distribute the ethanol to refineries to mix with gasoline.

Michael nods. "Bio-oil cannot be mixed with gasoline like ethanol. It has only half the energy of petroleum-based products. Scientists are trying to figure out how to modify the bio-oil to make a more useable product."

Bio-oil sounds closer to gasoline than ethanol, but it does not fit into our current energy production. I flip back through my notes. The best new fuels use existing oil refineries, pipelines, and gas stations.

I guess what Michael is trying to tell me is that the compounds created by the different processes are extremely complicated. Every compound is different and requires more research, but cutting-edge research is being done every day. I thank Michael and Sarah. I'm looking forward to learning more about other types of biofuels.

The combustion of gasoline in a car engine creates the energy that powers the car. Combustion is the process of burning something.

CHAPTER 7

The Power of Pond Scum

I **want to learn more about other types of biofuels. I read an article about more than 100,000 different kinds of red, green, and brown algae that grow all over the world. They range in size from huge seaweeds to microscopic organisms floating on the water's surface. Out of all the available algae, the pond scum on the water's surface is the best for biofuels because of its oil content. I imagine my dad pumping pond scum into his car. Algae seem about as far from biofuel as we can get. I am off to Arizona to get the whole scoop.**

I meet with Professor Marcus James at a university where algae grow in wall-sized tanks. "Algae are not fussy creatures," he says. "All they need are water, sunlight, and carbon dioxide to grow."

"Do they use photosynthesis?" I ask.

Someday the algae from this pond could be converted into fuel to power your car.

"Yes," Professor James says. "Although the temperature of the water is important, algae don't care if the water is fresh, salty, or brackish—a mixture of fresh and salt water. Algae love carbon dioxide, which is why some algae farms are built next to factories that produce a lot of carbon dioxide."

I think back to my visits to Trent's cornfield and Sarah's switchgrass field. Two fields, two different crops, but one product: ethanol.

"Is algae converted to ethanol, too?" I ask.

Professor James shakes his head. "Nope. We turn algae into biodiesel. Some algae contain fat. We can press about 75 percent of the fat out, like squeezing juice from an orange. The rest can be extracted with a special chemical. The fatty oil extracted from the algae is refined and mixed with catalysts to create biodiesel."

Algae are hardy organisms. They can grow almost anywhere there is light, water, and carbon dioxide. You might see algae growing at your local aquarium!

"Is biodiesel better than ethanol?" I ask.

"It could be," Professor James says. "Right now, ethanol is the only biofuel produced on a large scale. Algae-based fuel costs more than regular gasoline because of algae's limited availability. Plus, we're still trying to figure out the best time to grow algae, how much water they need, and when to harvest algae for the best oil production. Also, bacteria and other pests can contaminate algae pools, so we need to protect our crop."

Professor James is a fast talker. I write as fast as I can to keep up.

"Yet, even with these challenges, algae research continues," he says. "Algae's oil content makes it an attractive alternative as a drop-in replacement solution for petroleum-based fuel."

I remember what Mrs. Baker said. He means algae could be produced and transported using existing structures.

"An algae-growing biodiesel factory could produce up to 10 million gallons (38 million L) of biodiesel every year," Professor James says.

"Will that be enough to replace our dependence on petroleum?"

Professor James shakes his head. "Unfortunately, not even close. We would need to produce about 140 billion gallons (530 billion L) per year."

"Yikes. So biofuel made from algae is a long way off, then?"

"Not necessarily," Professor James says. "New facilities are in the works to test the production of algae on a large scale. Today, some airlines are experimenting with a blend of petroleum-based jet fuel and biofuel from algae oil. As long as researchers keep working and scientists keep experimenting, algae-based fuel will keep improving."

PIGS FEAST ON ALGAE

Scientists at Cornell University in Ithaca, New York, replaced soybean meal in livestock feed with the algae left over from biofuel production. Once the fat is harvested from algae for biofuel, scientists are left with no-fat algae that make for very healthy food. Although algae may not sound appetizing to us, pigs and chickens love them. This practice also transforms biofuel waste into valuable food for farm animals. Perhaps it could also free up valuable farmland for the production of human food.

CHAPTER 8

The Future of Biofuels

The last stop on my journey is a biofuels research center at a major university. Dr. Wilson greets me in her lab. (She is a Ph.D. doctor, not a medical doctor.) Her research focuses on engineering bacteria to make them convert as much sugar as possible into ethanol.

"If we can get bacteria to make more fuel faster, then biofuels could be less expensive than gasoline," Dr. Wilson says. "If biofuels become affordable to produce, then everyone wins: the manufacturers, the consumers, and the environment."

"Why do you engineer the bacteria?" I ask. "Don't they munch sugar on their own?"

Dr. Wilson nods. "But we want bacteria to work as efficiently as possible. Think of a train approaching a fork in the tracks. The train rides along the main line, but when a switch is thrown the train veers left. The switch changes the path of the train's energy, right? When we engineer bacteria, we change the path of their energy. In the case of biofuel production, we want as much of the bacteria's energy as possible focused on producing fuel."

Dr. Wilson tells me that for years and years the oil industry has been studying how to improve petroleum-based fuel. But research into biofuels is relatively new.

"Scientists all over the world are asking questions about biofuels," she says. "Some study proteins, which include complex substances such as enzymes. Proteins are necessary to life on Earth, but they can also be engineered to be used as a source of food energy for biofuel production. Proteins are one of the most plentiful groups of organic molecules on Earth. If we were able to use proteins instead of sugar as the food source

Biofuel technology continues to improve as scientists and innovators make new discoveries.

for microbes, we would bypass some of the problems with growing and processing grasses and algae. Microbes munch on the raw proteins and produce fuel. These scientists are now trying to figure out how to produce large quantities of protein-fed biofuel."

TOOLS FOR A MICROSCOPIC WORLD

To engineer bacteria, scientists must figure out the genetic makeup of the bacteria. Scientists identify and study the individual genes and the enzymes they create. Next, scientists figure out how to improve the way bacteria convert sugar to fuel. Scientists can remove one gene and strengthen another to give bacteria super powers. Simple tools like microscopes, flasks, and petri dishes are common in biofuel labs. Additionally, scientists use complex instruments that can copy and splice together genes and instruments that can precisely measure the ingredients of biofuel compounds.

Dr. Wilson expects that we'll be hearing more about new plants designed for creating biofuels. For instance, grasses that yield more sugar per pound would be especially useful. More sugar means more ethanol. Researchers continue to work on refining biofuel to strip molecules that are not consistent with high-energy fuel. She says I should watch for biofuels to begin replacing petroleum in the production of plastic or nylon.

Dr. Wilson swirls a small bottle of biofuel made by her super bacteria.

"That doesn't look like much fuel," I say.

Biofuels may help us balance our energy needs with the needs of our natural environment.

"It's not," Dr. Wilson says. "In the lab, we might make one liter of fuel at a time. An industrial scale is much larger. Once our super bacteria make a fuel we like, we will have to figure out how to make many thousands of liters at a time."

"At a price we can afford," I say.

"Exactly!" Dr. Wilson says.

CHAPTER 9

Your Turn

You have had a chance to follow Esmeralda as she conducted her research. Now it's time to think about what you learned. To meet our energy needs, biofuel must be inexpensive, environmentally friendly, and able to drop in to our current energy systems. Corn grain is easy to grow and extract the sugars from to make the ethanol, but it uses prime farmland that could produce food. Purpose-grown grasses such as switchgrass do not compete with our food supply, but the sugar is more difficult and expensive to access. Algae have the potential to create a great and very eco-friendly biofuel, but right now we could not produce nearly enough algae-based biofuel to meet our energy needs. While biofuel research has come a long way, scientists have not yet found the perfect biofuel. Maybe you'll take the next big step in biofuel research!

YOU DECIDE

1. Do you think the pros of corn ethanol outweigh the cons? Why?

2. Take a look around your house. Name some examples of biomass you see.

3. Current research focuses on engineering bacteria and proteins to keep fuel-producing microbes happy. How would you determine if one research project is more promising than another?

4. If you had $5 million to invest in biofuel research, which biofuel would you invest in? Why?

5. How big of a role do you think biofuels will play in the future of energy? Why?

Biofuel research is an ever-expanding science—maybe you'll make the next big discovery!

GLOSSARY

algae: Single- or multi-celled organisms that live in the water and carry out photosynthesis.

bacteria: Single-celled organisms that can only be seen with a microscope.

biodiesel: A clean-burning alternative fuel made from biomass.

biogas: A gas produced by the breakdown of organic material.

biomass: Organic material made from plants and animals.

butanol: An alcohol produced during the breakdown of sugars.

carbohydrates: A group of organic compounds that contain carbon, oxygen, and hydrogen molecules.

catalyst: A substance that speeds up a reaction.

cellulose: A carbohydrate found in the cell walls of green plants that gives a plant its strength and rigid structure.

enzyme: A protein that speeds up a chemical reaction in a living organism.

ethanol: An alcohol produced from corn or other biomass crops.

hydrocarbon: An organic compound made of carbon and hydrogen.

microbe: Living microscopic organisms, such as bacteria (E. coli) or fungi (yeast).

molecule: The smallest unit of a substance that still possesses the properties of that substance.

organic: Material from living plants or animals.

petroleum: A fossil fuel often called crude oil or oil.

photosynthesis: A process through which carbon dioxide, water, and light energy are converted to food for plants.

protein: A large group of complex substances necessary to life on Earth.

renewable: When something can be replaced by natural cycles in nature or the environment.

respire: To take in oxygen and give off carbon dioxide.

Does Biomass Give Off Heat?

Take a walk around your yard with an old coffee can. Collect grass clippings, leaves, and fallen flower petals. Add a few drops of water and stir around in the can until the biomass is damp. Cut an X in the middle of the cover. Place the cover over the can. Read a thermometer and record the room temperature. Then insert the thermometer through the cover and into the biomass. Check the temperature of the biomass immediately. Record the temperature on a graph. Record the time and the temperature every hour for a week. What do you notice about the temperature? Explain what is happening using the concepts you learned in this book.

Connect the Dots

Renewable energy is about connecting systems, for instance, using the waste of one process to fuel another. One example in this book refers to a dairy farm that recycles its manure to provide its own electricity. Can you find other examples of connected systems in your area? What are they? How do they connect? What types of renewable energy do they provide? Create a poster showing the relationship between the energy systems.

Shrink Your Carbon Footprint

The amount of greenhouse gas you produce is sometimes called your carbon footprint. Visit an online carbon footprint calculator to estimate how much carbon dioxide your household produces in a year. Examine your results—where can you reduce emissions? Can you hang laundry in the sun instead of running the clothes dryer? Can you replace outdoor lighting with solar-powered lamps? What about growing your own food to reduce driving trips to the grocery store? What are other things you can do to reduce emissions?

Growing your own food can be a great way to cut down on car trips.

SELECTED BIBLIOGRAPHY

"Algal-Based Biofuels and Biomaterials." *Arizona State University*, n.d. Web. Accessed May 12, 2012.

Beller, Dr. Harry R. Personal interview. Accessed May 8, 2012.

Edwards, Mark. "Algae 101: The Tiny Plant That Saved Our Planet." *Algae Industry Magazine*, April 14, 2012. Web. Accessed May 2, 2012.

Lane, Jim. "The State of the 2012 Advanced Biofuels Industry." *RenewableEnergyWorld.com*, March 15, 2012. Web. Accessed June 28, 2012.

"The Biofuels Technology Square Dance." *Biofuels Digest*, April 18, 2011. Web. Accessed May 8, 2012.

FURTHER INFORMATION

Books

Doeden, Matt. *Green Energy: Crucial Gains or Economic Strains?* Minneapolis: Twenty-First Century Books, 2010.

Morris, Neil. *Biomass Power*. North Mankato, MN: Smart Apple Media, 2007.

Thomas, Isabel. *The Pros and Cons of Biomass*. New York: The Rosen Publishing Group, 2008.

Websites

http://environment.nationalgeographic.com/environment/global-warming/biofuel-profile/
Start your biofuels research here with activities and links to other renewable energy resources.

http://www.eia.gov/kids/energy.cfm?page=biomass_home-basics-k.cfm#top-container
Visit this site for biomass facts, figures, pictures, and activities to share with your friends.

http://www.nsf.gov/news/special_reports/science_nation/greengasoline.jsp
Watch how green gasoline is made on this National Science Foundation video.

INDEX